PLANT GATEWAY'S
# THE GLOBAL FLORA
A practical flora to vascular plant species of the world

ANGIOSPERMS

## 1. AMBORELLACEAE

by

J.W. BYNG
&
M.J.M. CHRISTENHUSZ

January 2018

# The Global Flora

A practical flora to plant species of the world
Angiosperms, Amborellaceae Vol 3: 1-20.

Published by Plant Gateway Ltd., 5 Baddeley Gardens, Bradford, BD10 8JL, United Kingdom

© Plant Gateway 2018

This work is in copyright. Subject to statutory exception and to the provision of relevant collective licensing agreements, no reproduction of any part may take place without the written permission of Plant Gateway Ltd.

ISSN   2398-6336
eISSN 2398-6344
ISBN 978-0-9929993-7-7

Plant Gateway has no responsibility for the persistence or accuracy of URLS for external or third-party internet websites referred to in this work, and does not guarantee that any content on such websites is, or will remain, accurate or appropriate.

British Library Cataloguing in Publication data
A Catalogue record of this book is available from the British Library

For information or to purchase other Plant Gateway titles please visit
www.plantgateway.com

*Authors*

James W. Byng, Plant Gateway, Bradford & Kingston, United Kingdom and Den Haag, the Netherlands; Naturalis Biodiversity Center, Leiden, The Netherlands.

Maarten J.M. Christenhusz, Plant Gateway, Bradford & Kingston, United Kingdom and Den Haag, the Netherlands; Royal Botanic Gardens, Kew, United Kingdom.

Cover image: © Mike Bayly / CC BY-SA 3.0

## Summary

Amborellaceae is endemic to New Caledonia and contains one genus. The sole species in this genus, *Amborella trichopoda*, was originally described in Monimiaceae and following molecular studies was found to be sister to all other extant angiosperms. Diagnostic macro-morphological characters for the family are: dioecious trees or shrubs with simple, alternate and exstipulate leaves, axillary inflorescences, undifferentiated and spirally arranged perianths, superior ovaries, free carpels, and red drupaceous fruits. The wood lacks oil cells, vessels and grouped axial parenchyma cells, and has variation in ray seriates. Pollen grains are oblate to spheroidal, monoaperturate, with finely papillate ornamentation in the non-apertural region and a operculum forming in the apertural region. The sole species in the family is illustrated and comes with a description including data on its habitat, distribution, known herbarium specimens at major herbaria, cultivation requirements and additional observations.

# AMBORELLACEAE Pichon (1948: 384)

*Trees to small shrubs*, **dioecious**. *Stems* **lacking wood vessels**. *Leaves* simple, alternate, margins sinuate; petiolate; **estipulate**. *Inflorescences* **axillary**, borne in the axils of the leaves and on leafless branches, bracteate. *Flowers* functionally unisexual, actinomorphic, with a pedicellate receptacle. *Perianth* **undifferentiated**, spirally arranged, slightly fused basally; bracteoles spiralling on the receptacle and sometimes grading into tepals. *Male flowers:* stamens spiralling on the receptacle, more or less laminar, filaments short to absent; anthers introrse, dehiscing by longitudinal slits. *Pollen:* oblate to sphaeroidal, monoaperturate. *Female flowers:* ovary **superior**, carpels free and borne in the centre of the receptacle, stipitate; stigma sessile; locules 1, ovule 1, placentation marginal. *Fruits:* **fruiting carpels drupaceous**. *Seed* 1.

The family contains one genus and one species: *Amborella trichopoda* Baill.

**Distribution** – Endemic to New Caledonia (Figure 1).

**Classification** – The genus was originally named and described by Baillon (1869) in a footnote with additional notes added later by Baillon (1873). The species was placed in Monimiaceae because of its similarity to the flowers of *Hedycarya* J.R.Forst. & G.Forst. before being placed in its own family by Pichon (1948). Molecular studies in the last two decades have shown the genus to be the most basal lineage of extant angiosperms. In APG classifications *Amborella* is a monotypic family placed in its own order (APG IV, 2016; Figure 2).

Figure 1: Global distribution of Amborellaceae (highlighted).

Figure 2: Classification of Amborellaceae showing the order and generic rank relevant to the family. Adapted from APG IV (2016).

**Wood anatomy** (Figure 3) – Growth ring boundaries are indistinct. Wood vessels are absent. Tracheids are the abundant cell type in wood with many conspicuously bordered pits (7-10 µm in diameter, often arranged uniseriately) in mainly the radial and sporadically in the tangential walls. Mean tracheid length is 2850–3140 µm (up to 4660 µm in the outer parts of mature wood). Apotracheal axial parenchyma is scanty diffuse, with about five cells per parenchyma strand. Rays in the first formed secondary xylem are uni- or biseriate with mainly erect cells, but multiseriate rays (3–5 cells in width) with a higher proportion of square and procumbent cell shapes become abundant in mature wood. Ray cells do not contain crystals, silica bodies, ethereal oils or mucilage.

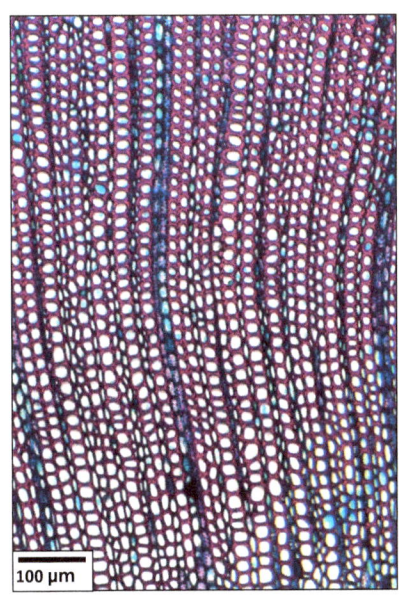

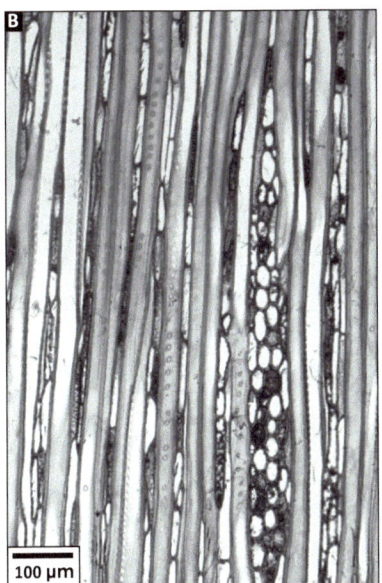

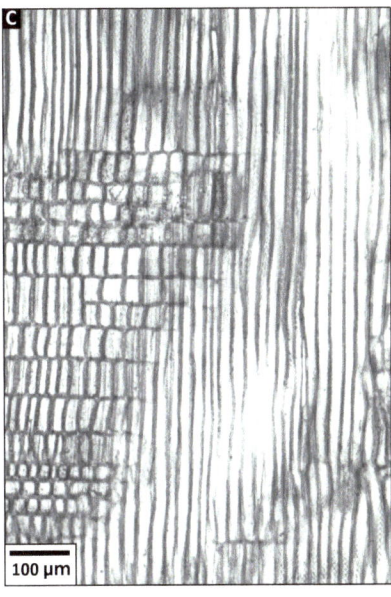

Figure 3: Wood anatomy of *Amborella trichopoda*. A, transverse section showing vesselless xylem; B, tangential longitudinal section showing a mix of uniseriate and multiseriate rays; C, radial longitudinal section of a juvenile stem showing a high proportion of upright and square cells. Images: A, © Frederic Lens; B-C, © Jugo Ilic.

**Pollen morphology** – The pollen of *Amborella* (Figure 4) is shed as small-sized monads (about 10–25 μm), ulcerate, with heteropolar polarity. Pollen grain shape is oblate to spheroidal in hydrated condition or irregularly shaped and folded in dry condition. The pollen is monoaperturate, with the aperture margin poorly defined and a operculum forming in the apertural region. The ornamentation is finely papillate in the non-apertural region.

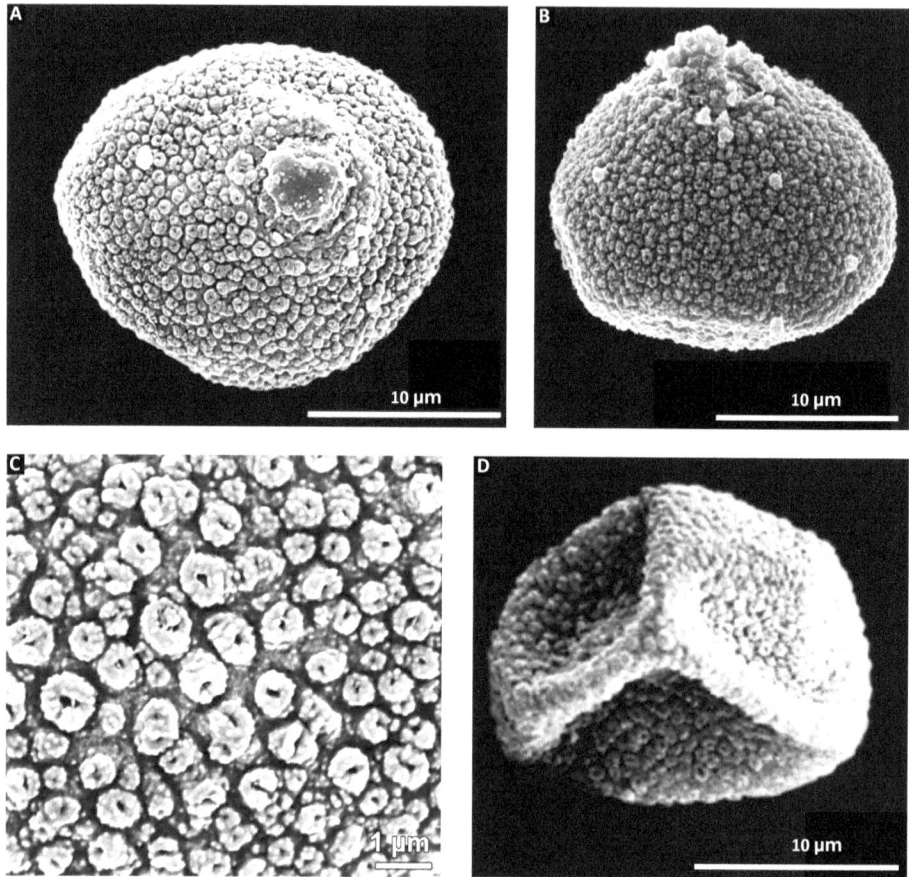

Figure 4: Pollen morphology of *Amborella trichopoda*. A, equatorial view; B, polar view; C, detail of ornamentation in the non-apertual region; D, dry pollen. © Heidemarie Halbritter.

**References** – APG IV (2016), Bailey (1957), Bailey & Swamy (1948), Baillon (1869, 1873), Byng (2014), Carlquist & Schneider (2001), Halbritter (2018), Hesse (2001), Jérémie (1982), Jérémie *et al.* (2008), Perkins (1925), Philipson (1993), Pichon (1948), Qiu *et al.* (1999), Sampson (1993).

# 1. *Amborella* Baillon (1869: 328)

**Type** – *Amborella trichopoda* Baill.

**Etymology** – *Amborella* is the diminutive form of 'ambora', a native Malagasy plant name for *Tambourissa* Sonn. (Monimiaceae), to which *Amborella* was originally thought to be related.

Description as for the family.

## 1. *Amborella trichopoda* Baillon (1869: 328)

Figures 5, 6, 7, 8, 9. Table 1.

**Type** – NEW CALEDONIA. Mt. Arago, 27 November 1869, *B. Balansa 1800* (lectotype designated by Jérémie, 1982): P [P00181299]; isolectotypes BM [BM001209670], GH [GH00039909], P [P00181300]). Remaining syntypes – NEW CALEDONIA. Balade, *E. Vieillard 32* (P [P00181370, P00181371, P00181372]); NEW CALEDONIA. Wagap, *E. Vieillard 2296* (P [P00181367, P00181368, P00181369]).

**Etymology** – trichopoda, from Greek 'thrix', hair and pous, foot, referring to the often hairy pedicels and petioles.

*Tree or shrub,* 2–6 m tall. *Stems* cylindrical. *Leaves* dark green, often shiny above, light green below, chartaceous; petiole 6–12 mm long. Blades 6.0–22.5 × 2.0–8.5 cm, ovate to ovate-lanceolate, base rounded or truncate to subcordate, apex rounded to acute, often apiculate; densely pubescent to glabrous; **margins variable ranging from undulate to toothed** (6–12 teeth per side); secondary nerves 6–10 pairs, generally visible below. *Inflorescences* cymose with 2–30 flowers, 1.5–6.0 cm long, numerous inflorescences can arise from one leaf axil. *Flowers* ca. 3–5 mm in diameter at anthesis, open during day and night; **unisexual**; pedicel 2–12 mm long; tepals 4–8, ca. 2 mm long, **whitish to slightly greenish**. *Male flowers:* white to pale yellowish-green; stamens 6–25, **sessile**, 1.5–1.8 mm long, triangular; filament broadly flat, numerous unicellular hairs at base, greenish; outer members larger, inserted in a spiral on the receptacle with outer members attached to the base of perianth; anthers pointed with four pollen sacs in two thecae, whitish. *Female flowers:* slightly larger than male flowers, white to pale yellowish-

Figure 5: Images of living *Amborella trichopoda*. A, close-up of male flower; B, male branch showing axillary inflorecenses; C, female inflorescence; D, close-up of female inflorescence; E, immature fruits; F, ripe fruits. Images: A, © Mike Bayly / CC BY-SA 3.0; B, © Scott Zona/ CC-BY-2.0; C & F, © Rogier van Vugt; D & F © Maarten Christenhusz.

Figure 6: Herbarium specimens of *Amborella trichopoda*. A, upper leaf surface (syntype: *Pancher s.n.*); B, lower leaf surface (lectotype: *Balansa 1800*); C & D, upper and lower surfaces with enlarged leaf tooth inset (*Byng 200*; cultivated at *Hortus Botanicus*, Leiden); E, fruits (*MacKee 5181*). Images original.

green; carpels 3–8, free, pitcher-shaped, ca. 2 mm long, greenish; **mostly 1–2 staminodes**, whitish, often a rudimentary anther; ovule pendulous, sessile, anatropous; stigma oblique, with two conspicuous feathery flanges, white to yellow. *Fruits* 8–10 × 4–8 mm, obovoid with slightly compressed sides and apex, green when immature and **red when mature**. *Seed* ca. 8 mm long, embryo minute.

Figure 7: Herbarium specimens of *Amborella trichopoda*. A, male inflorescence with enlarged flower inset (*MacKee 26532*); B, female inflorescence with enlarged flower inset (*MacKee 26557*; S = staminode, C = carpel). Images original.

**Distribution** – 60: NWC. The species is found across the central part of Grande Terre, from Dogny - Mt Canala to Mt Tonine (Touho). The localities where the species is most abundant are Sarraméa, Tchamba and Touho.

**Habitat and ecology** – Dense wet evergreen forests on schistose soils; 80–1000 m elevation

**Phenology** – Flowers lack scent and open between January and June. Fruits develop and ripen irregularly throughout the year.

**Pollinators** – The species is ambophilous with both insects and wind known to be effective pollinators.

**Chromosome size** – 2n = 26 (Oginuma *et al.*, 2000).

**Genome size** – 1C = 0.89 pg (Leitch & Hanson, 2002).

**Cultivation** – *Amborella* is not easily kept in cultivation. The two main factors affecting successful cultivation of the species are temperature and fungal infections. The species prefers to grow in the shade and thrives best in temperatures between 16–18°C with a minimum temperature of 10°C. Temperatures higher than 25°C should be avoided because they increase the occurrence of the oomycete pest *Pythium splendens* Braun. This fungal infection is not dangerous for the plant unless temperatures are very high (above 35°C). The plant can be treated with the fungicide Aliette®. Good water quality is also of importance, as the species does not grow in water that is too alkaline. Vegetative propagation is possible but 2–3 leaf pairs are usually needed and the freshly cut stem (and stem of original plant) should be treated with charcoal powder immediately after cutting to prevent disease, followed by adding a rooting hormone containing 3-indole butyric acid before insertion into the compost. The ideal compost used is a 60% peat-based mixture (70% peat and 30% clay dust, pH 5.8), 20% pumice and 20% lava, with fertiliser added fortnightly or weekly in higher temperatures (Große-Veldman *et al.*, 2011).

**Preliminary conservation status** – Least concern (LC) according to the IUCN Red List Criteria (IUCN, 2014), as the species is locally abundant and faces no immediate threats.

**References** – Bailey & Swamy (1948), Baillon (1869), Endress & Igersheim (2000), Große-Veldman *et al.* (2011), IUCN (2014), Jérémie (1982), Jérémie *et al.* (2008), Leitch & Hanson (2002), Oginuma *et al.* (2000), Thien *et al.* (2003).

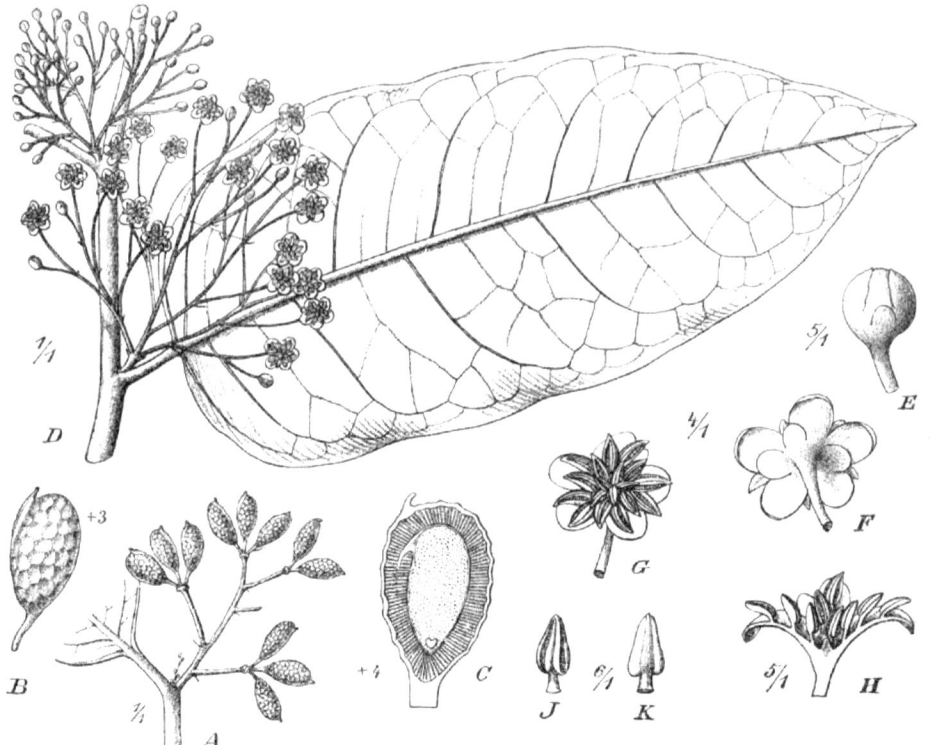

Figure 8: Illustration of *Amborella trichopoda*. A, fruiting branch; B, close-up of fruit; C, longitudinal section of fruit; D, branch with male flowers and leaf; E, male flower in bud; F, abaxial side of male flower; G, adaxial side of male flower; H, longitudinal section of male flower; J & K, stamen. Figure from Perkins (1911).

Table 1: Herbarium specimens consulted at BM, K, L and P of *Amborella trichopoda*. Duplicates and specimens at other herbaria are listed where known.

| Collector | Collector Number | Collection Date | Locality | Herbarium | Phenology | Altitude (m) |
|---|---|---|---|---|---|---|
| Balansa, B. | 1800 | 27/11/1869 | Mt. Arago | BM, GH, P | Fr | 700 |
| Bernardi, L. | 9959 | 03/08/1965 | Koindé | G, K, L, P | Fr | 900 |
| Blanchon, J.P. | 983 | | Forêt au pied du Mé-Aoui | NOU, P | Fr | 600 |
| Blanchon, J.P. | 1009 | | Plateau-Col des Roussettes, route forestière | NOU, P | Fr | |
| Bradford, J.C. et al. | 1172 | 26/11/2002 | Ascent of Plateau de Dogny from Sarramea | K, MO, NOU | Fr | 760 |
| Carlquist, S. | 15333 | 02/08/1977 | De Sarraméa au sommet du Plateau de Dogny | K, L, NOU, P | Fr | 80 |
| Christenhusz, M.J.M. | 6192 | 02/03/2011 | Plateau de Tango, Riviere Morlet | H | Sterile | |
| Coulerie, P. | 27 | 15/05/2009 | Plateau de Dogny | NOU | Fl Fr | 646 |
| Dagostini, G. | 162 | 24/10/1999 | Piste du Dogny | K, NOU, P | Fr | 300-800 |
| Dickison, W.C. | 264 | 16/12/1981 | Katrikoin, road to MéOri, above Katrikoin | NOU | Fr | |
| Douglas, S. | 35 | 16/11/2005 | Plateau de Dogny | NOU | Fr | 850 |
| Duretto, M.F. | 638 | 25/09/1995 | Track to Plateau de Dogny, from Sarramea | MEL | Fr | |
| Duretto, M.F. | 639 | 25/09/1995 | Track to Plateau de Dogny, from Sarramea | MEL | Fr | |
| Edmondson, J.R. | 3752 | 11/08/1981 | Forest road ascending from Col Toma to Table Unio | K | Fr | 800 |
| Grignon, C. | 99 | 14/01/2009 | Réserve du Col d'Amieu-rivière Fa Foméc hawa | NOU, P | Fl Fr | |
| Hoff, M. | 2688 | 29/09/1980 | Plateau de Dogny | NOU | Fr | 600 |
| Hoff, M. | 3247 | 09/12/1980 | Aoupinié | K, NOU, P | Fr | 950 |
| Jaffré, T. | 2938 | 24/03/1988 | Exploitation forestière Guiraud, piste en direction du Menazi (Bokoua) | NOU, P | Fl | 600 |
| Jérémie, J. | 1587 | 09/01/1987 | Mt. Dogny | MO, NOU, P | Fr | 800 |
| Lécard, T. | 23 ? | | | P | | |
| Lécard, T. | | 20/10/1879 | Forêt de Bourail | P | Fr | |
| Lorence, D.H. | 8346 | 28/06/1998 | Trail to Plateau de Dogny from Hotel Evasion 130. | BISH, K, MO, NY, P, US | Fr | 640-900 |
| Lorence, D.H. | 8349 | 28/06/1998 | Trail to Plateau de Dogny from Hotel Evasion 130 | AD, MO, US | Fl | 850 |
| Lowry II, P.P. | 3586 | 09/11/1984 | Plateau de Dogny | MO, NOU, P | Fr | 680 |
| Lowry II, P.P. | 4725 | 10/12/1996 | Plateau de Dogny | MO, P | Fr | 870 |
| Lowry II, P.P. | 5569 | 09/03/2002 | Plateau de Dogny | MO, NOU, P | Fl | 515 |
| MacKee, H.S. | 5181 | 29/08/1956 | Au-dessus Atéou : Crête entre Kamendoua et Tipindjé Mt. Pouitchaté | E, K, L, P | Fr | 700-1000 |
| MacKee, H.S. | 5182 | 29/08/1956 | Au-dessus Atéou : Crête entre Kamendoua et Tipindjé Mt. Pouitchaté | E, L, P | Sterile | 700-1000 |
| MacKee, H.S. | 5591 | 24/10/1956 | Canala : Creek à 4 km Kouaoua-La Foa | L, P | Sterile | 100 |
| MacKee, H.S. | 5617 | 25/10/1956 | La Foa : Pente W Plateau Dogny | K, L, P | Fr | 300-900 |

Table 1 continued

| | | | | | | | |
|---|---|---|---|---|---|---|---|
| MacKee, H.S. | 5618 | 25/10/1956 | La Foa : Pente W Plateau Dogny | L, P | | 300-900 | |
| MacKee, H.S. | 10028 | 16/01/1963 | Ponérihouen : Expl. For. Devillers au-dessus Nohéa | K, L, NSW, P | Fr | 400-500 | |
| MacKee, H.S. | 12148 | 07/02/1965 | Haute Vallée La Foa : Contrefort Mt. Nemmara | K, L, NOU, P | Fr | 500-650 | |
| MacKee, H.S. | 12322 | 30/03/1965 | Pente W Plateau Dogny | BISH, CANB, K, L, MO, NOU, P, U | Fl | 700 | |
| MacKee, H.S. | 12323 | 30/03/1965 | Pente W Plateau Dogny | CANB, K, P | Fl | 700 | |
| MacKee, H.S. | 12344 | 01/04/1965 | 20 km à l'E Col des Roussettes | BISH, K, L, MO, NSW, NOU, P | Fl | 500 | |
| MacKee, H.S. | 12582 | 10/05/1965 | Contrefort S Table Unio | P | Fr | 800 | |
| MacKee, H.S. | 12803 | 14/06/1965 | Expl. For. Launay Katrikoin | P | Fr | 500 | |
| MacKee, H.S. | 12832 | 30/06/1965 | Mé Ouié : Haute/ Crête Haute/ Boghen Kouaoua | P | Fr | 700 | |
| MacKee, H.S. | 15030 | 20/05/1966 | Col d'Amieu: Versant Couli | BISH, K, L, NSW, NOU, P | Fl Fr | 400 | |
| MacKee, H.S. | 16762 | 11/05/1967 | Haute Tchamba: Expl. For. Létocart | K, L, NOU, P | Fl | 300-500 | |
| MacKee, H.S. | 18055 | 24/11/1967 | Ponérihouen : Expl. For. Devillers - Pente Nord-Est du Mt. Aoupinié | K, P | Fr | 400-500 | |
| MacKee, H.S. | 18082 | 05/12/1967 | Canala : Sentier Ciu-Koindé | K, P | Fr | 500 | |
| MacKee, H.S. | 18703 | 28/04/1968 | Expl. For. Létocart Haute Vallée Amoa | P | Fl | 300 | |
| MacKee, H.S. | 18732 | 29/04/1968 | Haute Tchamba: Expl. For. Létocart | K, L, NSW, NOU, P | Fl Fr | 550 | |
| MacKee, H.S. | 23887 | 28/06/1971 | Contrefort S Mt. Canala : Col Ema-Koindé | K, L, NOU, P | Fr | 800-900 | |
| MacKee, H.S. | 23964 | 18/07/1971 | Houaïlou : Vallée Ba | BISH, K, L, NSW, P | Fr | 200-300 | |
| MacKee, H.S. | 26404 | 16/03/1973 | Touho : Pente N Tonine | NSW, P | Fl | 600-800 | |
| MacKee, H.S. | 26412 | 16/03/1973 | Touho : Pente N Tonine | K, L, P | Fl | 800-1000 | |
| MacKee, H.S. | 26532 | 11/04/1973 | Col Amieu : Mt. Pembaï | BISH, K, L, MO, NSW, P | Fl | 600-800 | |
| MacKee, H.S. | 26557 | 11/04/1973 | La Foa : Col Ciu-Coindé | K, L, P | Fl | 800 | |
| MacKee, H.S. | 26727 | 15/05/1973 | Touho : Ponandou | K, L, P | Fl | 30-100 | |
| MacKee, H.S. | 28416 | 30/03/1974 | Tiwaka : Pente S Inédèté | P | Fr | 450 | |
| MacKee, H.S. | 28460 | 31/03/1974 | Tiwaka : Pente E Moindip | MO, NOU, P | Fr | 700 | |
| MacKee, H.S. | 29032 | 27/07/1974 | Ponérihouen : Expl. For. Devillers Pente E Mt. Aoupinié | K, P | Fr | 500 | |
| MacKee, H.S. | 31574 | 14/07/1976 | Ponérihouen : hauteurs de Nato (crête Népia-Tchamba) | K, L, P | | 400 | |
| MacKee, H.S. | 31951 | 11/09/1976 | Touho : Tipouatène | P | | 500 | |
| MacKee, H.S. | 38408 | 04/12/1980 | Pente S Mt. Yora | P | | 700 | |
| MacKee, H.S. | 38907 | 07/04/1981 | Haute Vallée Mou | K, NOU, P | Fr | 600 | |
| MacKee, H.S. | 38909 | 07/04/1981 | Haute Vallée Mou | K, L, NOU, P | Fl | 600 | |
| MacKee, H.S. | 41333 | 21/03/1983 | Contrefort W Mé Aoui | K, P | Fl | 500 | |
| MacKee, H.S. | 43303 | 14/10/1986 | Haute Néoua : Bokoua | L, NOU, P | Fr | 550 | |
| McPherson, G. | 2359 | 24/01/1980 | Plateau de Dogny | MO, P | Fr | 750 | |
| McPherson, G. | 2607 | 22/04/1980 | Plateau de Dogny | AK, BISH, CANB, K, L, MO, NOU, NSW, P | Fl Fr | 600 | |
| McPherson, G. | 4187 | 22/09/1981 | Mé Ori près Katrikoin | L, MO, NOU, NSW, P | Fr | 500 | |
| McPherson, G. | 5088 | 07/11/1982 | Mt. Rembai, S Col d'Amieu | MO, NOU, NSW, P | Fr | 500 | |

Table 1 continued

| | | | | | | |
|---|---|---|---|---|---|---|
| McPherson, G. | 5727 | 07/06/1983 | Mt. Rembai, au dessus du Col d'Amieu | MO, NOU, NSW, P | Fr | 550 |
| McPherson, G. | 5728 | 07/06/1983 | Mt. Rembai au dessus du Col d'Amieu | BISH, K, MO, NOU, NSW, P | Fr | 550 |
| McPherson, G. | 6297 | 02/02/1984 | Riv. Ponandou, S Touho | MO, NOU, P | Fl | 170 |
| McPherson, G. | 6521 | 07/05/1984 | Riv. Néaoua, S Houailou | MO, NOU, P | Fr | 450 |
| McPherson, G. | 6522 | 07/05/1984 | Riv. Néaoua, S Houailou | MO, P | Fl | 450 |
| McPherson, G. | 18188 | 12/04/2001 | Mt. Nakada | MO, NOU, P | Fr | 550 |
| McPherson, G. | 18189 | 12/04/2001 | Mt. Nakada | MO, P | Fl | 550 |
| McPherson, G. | 18521 | 24/04/2002 | Mt. Goro Até | MO, NOU, P | Fr | 890 |
| McPherson, G. | 18554 | 26/04/2002 | Upper Amoa River Valley, along road near Col de Néûni | L, MO, NOU, P | Fl | 350 |
| McPherson, G. | 18696 | 08/05/2002 | Ponandou River Valley | MO, NOU, P | Fr | 600 |
| Morat, P. | 5083 | 20/07/1976 | Mt. Aoupinié | NOU, P | Fr | 850 |
| Morat, P. | 6533 | 02/04/1980 | Nakada | K, NOU, P | Fl | 600 |
| Morat, P. | 7952 | 10/03/1988 | Dogny | NOU, P | Fl | 700 |
| Morat, P. | 7962 | 17/03/1988 | Aoupinié | NOU, P | Fl | 500 |
| Müller, I.H. | 41 | 09/02/1990 | Plateau de Dogny | P, Z | Fr | 250 |
| Munzinger, J.K. | 754 | 16/04/2001 | Mt. Nakada | MO, NOU, P | Fl | 780 |
| Munzinger, J.K. | 1411 | 10/11/2002 | Haute Tchamba | MO, NOU, P | Fr | |
| Munzinger, J. et al. | 2724 | 18/03/2004 | Col d'Amieu | NOU | Fertile | |
| Munzinger, J. et al. | 2725 | 18/03/2004 | Col d'Amieu | NOU | Fr | |
| Munzinger, J. et al. | 2726 | 18/03/2004 | Col d'Amieu | NOU | Fertile | |
| Munzinger, J. et al. | 2727 | 18/03/2004 | Col d'Amieu | NOU | Fertile | |
| Munzinger, J. et al. | 2728 | 18/03/2004 | Col d'Amieu | NOU | Fr | |
| Munzinger, J. et al. | 2729 | 18/03/2004 | Col d'Amieu | NOU | Fl | |
| Munzinger, J. et al. | 2730 | 18/03/2004 | Col d'Amieu | NOU | Fr | |
| Munzinger, J. et al. | 2731 | 18/03/2004 | Col d'Amieu | NOU | Fl | |
| Munzinger, J. et al. | 2732 | 18/03/2004 | Col d'Amieu | NOU | Sterile | |
| Munzinger, J. et al. | 2733 | 18/03/2004 | Col d'Amieu | NOU | Fr | |
| Munzinger, J.K. | 4789 | 15/11/2007 | Crête Aoupinié | MO, NOU, P | Fr | 900-1000 |
| Munzinger, J.K. | 6560 | 16/02/2011 | Tchamba, Grotiécou | NOU | Fr | 540 |
| Pancher, J. | | 1863 | | P | Fr | |
| Phillips, R.B. | 3193 | 26/07/1978 | Mt. Rembai, N Col d'Amieu | NOU, P | Fr | 830 |
| Pignal, M. | 2352 | 24/04/2004 | Vallée de la Tchamba | P | Fl | |
| Pillon, Y. et al. | 34 | 17/03/2005 | Mont Aoupiné, ancienne scierie Devilliers | NOU | Fl | |
| Pillon, Y. et al. | 35 | 16/03/2005 | Touho, rivière de la Ponandou, près des captages | NOU | Fl | |
| Pillon, Y. | 411 | 23/06/2006 | Tonine, flanc est | NOU | Fr | 480 |
| Pillon, Y. et al. | 1361 | 01/04/2009 | Crête au nord des tribus de Boregaou et de Bouirou, à l'est du Col des Roussettes. Bourai | NOU | Fl Fr | |
| Pillon, Y. et al. | 1363 | 07/04/2009 | Mé Ori, au-dessus de la tribu de Katrikoin. Moindou | NOU | Fr | |

Table 1 continued

| | | | | | | |
|---|---|---|---|---|---|---|
| Pillon, Y. et al. | 1379 | 08/04/2009 | Mé Foméchawa, au-dessus de la tribu de Katrikoin | NOU | Fl | |
| Pillon, Y. et al. | 1383 | 21/04/2009 | Mont Tonine, tribu de Paola, Touho | NOU | Fl Fr | 450 |
| Pillon, Y. et al. | 1395 | 23/04/2009 | Mont Pwicaté, tribu d'Atéu, Koné | NOU | Fr | 690 |
| Pillon, Y. et al. | 1406 | 29/04/2009 | Mont Nakada, tribu de Kouaré, Thio | NOU | Fl | 680 |
| Pillon, Y. et al. | 1409 | 05/05/2009 | Forêt de Neureu-éré, plateau au-dessus de la tribu de Ba, Houaïlou | NOU | Fl | 260 |
| Pillon, Y. & Boré, J.-M. | 1424 | 27/05/2009 | Col d'Amieu | MPU, NOU, SUVA, WAIK | Fl | 490 |
| Schmid, M. | 1212 | 04/1966 | Katrikoin | NOU, P | Fl | 600 |
| Schmid, M. | 3071 | 03/1970 | Mt. Kujua | L, NOU, P | Fl | 600 |
| Schmid, M. | 4626 | 22/04/1973 | Dogny | K, L, P | Fl | 800 |
| Schmid, M. | 5041 | 12/09/1974 | Montée de Dogny | NOU, P | Fr | 800 |
| Schodde, R. | 5245 | 17/01/1968 | Table Unio, pentes S | A, AD, CANB, L, P | Fr | 50 |
| Sevenet-Pusset | 1258 | 07/06/1977 | Dogny | NOU, P | | 450 |
| Simmons, M.P. | 1846 | 04/12/1997 | Plateau de Dogny | MO, P | Fr | 715 |
| Simmons, M.P. | 1860 | 09/12/1997 | Hte vallée de la Tchamba | MO, P | Fr | 540 |
| Snow, N. et al. | 9201 | 09/08/2003 | Mt. Aoupinié | AS, COLO, CS, GREE, NOU, P, RM, WELTU | Fr | 900 |
| Stone, B.C. | 14874 | 11/08/1981 | Mts above Kenerou, N of Sarramea, W of Dogny | NOU | Fr | |
| Suprin, B. | 839 | 05/11/1980 | Col d'Amieu, forêt de Rembaï | NOU | Fr | |
| Suprin, B. | 891 | 13/11/1980 | Col d'Amieu, forêt de Rembaï | NOU | Fr | |
| Thorne, R.F. | 28348 | 29/10/1959 | Plateau de Dogny | P | Fr | 800 |
| Tirel, C. | 1375 | 01/09/1978 | Vallée de la Tchamba | K, P | Fr | 400 |
| Tronchet, F. | 333 | 11/11/2002 | Vallée de la Haute Tchamba | MO, NOU, P | Fr | |
| Veillon, J.M. | 179 | 01/06/1965 | Col des Roussettes | NOU | Fr | 500 |
| Veillon, J.M. | 1212 | 13/04/1966 | Katrikoin | P | Fr | 600 |
| Veillon, J.M. | 2746 | 09/1972 | Versant E de l'Aoupinié | K, NOU, P | Fr | |
| Veillon, J.M. | 3273 | 24/08/1977 | Plateau de Dogny | K, NOU, P | Fr | 600-800 |
| Veillon, J.M. | 4630 | 20/11/1981 | Plateau de Tango : hte Tiwaka en direction de Grandié | K, NOU, P | Fr | 650 |
| Veillon, J.M. | 5441 | 12/04/1983 | Néaoua : près du site de la conduite | K, NOU, P | Fl | |
| Veillon, J.M. | 5696 | 16/05/1984 | Hte vallée de la Tchamba | NOU, P | Fl | 400 |
| Vieillard, E. | 32 | | Balade : bois humides des montagnes | P | Fl Fr | |
| Vieillard, E. | 2296 | | Wagap : bois des hautes montagnes | P | Fl Fr | |
| Vieillard, E. | 3149 | 1861-67 | Wagap | BM, K, P | Fl Fr | |
| Vieillard, E. | | | | P | Fr | |
| Waters, T. | 203 | 05/08/2001 | Plateau de Dogny, Sarraméa | FHO, P | | 700 |
| Weston, P.H. & Thien, L. | 2508 | 10/04/2001 | Track from Sarramea to Plateau de Dogny | NSW | | 800 |
| Weston, P.H. & Thien, L. | 2509 | 10/04/2001 | Track from Sarramea to Plateau de Dogny | NSW | | 800 |
| Weston, P.H. & Thien, L. | 2510 | 10/04/2001 | Track from Sarramea to Plateau de Dogny | NSW | Fl | 800 |
| Weston, P.H. & Thien, L. | 2511 | 10/04/2001 | Track from Sarramea to Plateau de Dogny | NSW | Fl Fr | 800 |
| Ziarnik, W.G. | 61 | 07/11/1982 | Mt. Rembaï | NOU | Fr | |

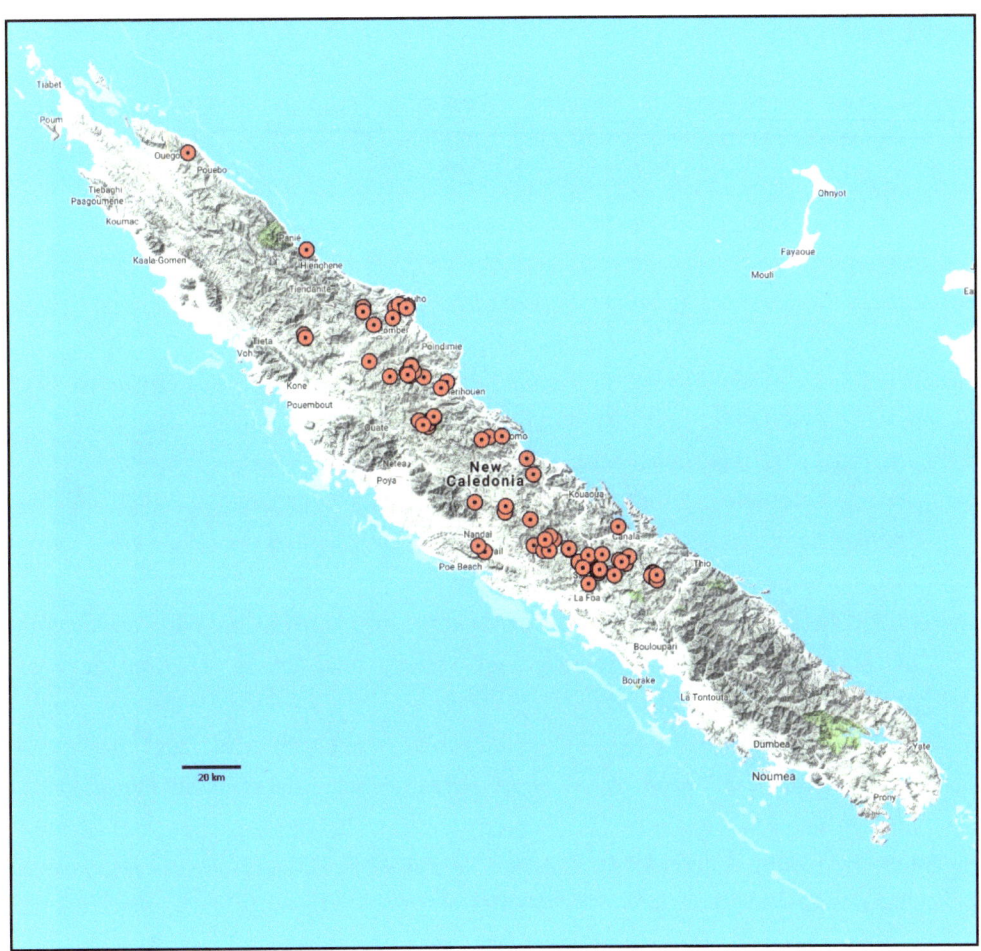

Figure 9: Distribution of *Amborella trichopoda* in New Caledonia based on herbarium records listed in Table 1.

# REFERENCES

APG IV. (2016) An update of the Angiosperm Phylogeny Group classification for the orders and families of flowering plants: APG IV. *Botanical Journal of the Linnean Society* 181: 1–20.

Bailey, I.W. (1957) Additional notes on the vesselless dicotyledon, *Amborella trichopoda* Baill. *Journal of the Arnold Arboretum* 38: 374–378.

Bailey, I.W. & Swamy, B.G.L. (1948) *Amborella trichopoda* Baill.: a new morphological type of vesselless dicotyledon. *Journal of the Arnold Arboretum* 29: 245–254.

Baillon, H.E. (1867–1869) *Histoire des Plantes*. Vol 1. Librairie de L. Hachette et Ciet, Paris.

Baillon, H.E. (1872–1873) *Histoire des Plantes*. Vol 4. Librairie de L. Hachette et Ciet, Paris.

Byng, J.W. (2014) *The flowering plants handbook*. Plant Gateway, Hertford.

Carlquist, S.L. & Schneider, E.L. (2001) Vegetative anatomy of the New Caledonian endemic *Amborella trichopoda*: relationships with the Illiciales and implications for vessel origin. *Pacific Science* 55: 305–312.

Endress, P.K. & Igersheim, A. (2000) The reproductive structures of the basal angiosperm *Amborella trichopoda* (Amborellaceae). *International Journal of Plant Sciences* 161(Suppl.): S237–S248.

Große-Veldman, B., Korotkova, N., Reinken, B., Lobin, W. & Barthlott W. (2011) *Amborella trichopoda* - Cultivation of the most ancestral angiosperm in botanic gardens. *Sibbaldia* 9: 143–155.

Halbritter, H. (2018) *Amborella trichopoda*. In: PalDat (2015-12-04) - a palynological database. Published on the Internet https://www.paldat.org/pub/Amborella_trichopoda/300223 [accessed 10 January 2018]

Hesse, M. (2001) Pollen Characters of *Amborella trichopoda* (Amborellaceae): a reinvestigation. *International Journal of Plant Sciences* 162: 201–208.

IUCN. (2014) *IUCN Red List Categories and Criteria*: Version 3.1. 2nd Edition. IUCN Species Survival Commission, IUCN Gland & Cambridge.

Jérémie, J. (1982) Amborellaceae. In: Aubréville, A. & Leroy, J.-F. (eds.). *Flore de la Nouvelle Calédonia et dépendances*. Vol 11. Musée National d'Histoire Naturelle, Paris.

Jérémie, J., Lowry, P.P. & Tronchet, F. (2008) Amborellaceae. *Species Plantarum: Flora of the World* 14: 1–7.

Leitch, I.J. & Hanson, L. (2002) DNA C-values in seven families fill phylogenetic gaps in the basal angiosperms. *Botanical Journal of the Linnean Society* 140: 175–179.

Oginuma, K., Jaffré, T. & Tobe, H. (2000) The karyotype analysis of somatic chromosomes in *Amborella trichopoda* (Amborellaceae). *Journal of Plant Research* 113: 281–283.

Perkins, J. (1911) Monimiaceae. In: Engler, H.G.A. (ed.). *Das Pflanzenreich*. Vol. 101. Verlag von Wilhelm Engelmann, Leipzig.

Perkins, J. (1925) *Übersicht über die Gattungen der Monimiaceae*. W. Engelmann, Leipzig.

Philipson, W.R. (1993) Amborellaceae. In: Kubitzki, K., Rohwer, J.G. & Bittrich, V. (eds.). *The families and genera of vascular plants.* 2: 92.

Pichon, P. (1948). Les Monimiacées, famille hétérogene. *Bulletin de la Muséum d'Histoire naturelle* 20: 383–384.

Qiu, Y.-L., Lee, J., Bernasconi-Quadroni, F., Soltis, D.E., Soltis, P.S., Zanis, M., Zimmer, E.A., Chen, Z., Savolainen, V. & Chase, M.W. (1999) The earliest angiosperms: evidence from mitochondrial, plastid and nuclear genomes. *Nature* 402: 404–407.

Sampson, F.B. (1993) Pollen morphology of the Amborellaceae and Hortoniaceae (Hortonioideae: Monimiaceae). *Grana* 32: 154–162.

Thien, L.B., Sage ,T.L., Jaffré, T., Bernhardt, P., Pontieri, V., Weston, P.H., Malloch, D., Azuma, H., Graham, S.W., McPherson, M.A., Rai, H.S., Sage, R.F. & Dupré, J.-L. (2003) The population structure and floral biology of *Amborella trichopoda* (Amborellaceae). *Annals of Missouri Botanic Garden* 90: 466–490.

## ACKNOWLEDGEMENTS

We thank the curators of BM, K, L and P for access to their collections. We also thank Jugo Ilic, Heidemarie Halbritter and Rogier van Vugt for allowing us to use their photos. Lastly, we thank Frederic Lens for his advice on wood anatomy and use of one of his images.

# INDEX TO SCIENTIFIC NAMES

*Amborella* Baill. . . . . . . . . . . . . . . . . . . . . . . . . . . . . . . . . . . . . . . . . . . . . . . . . . . . . . . . . . . . . . . . . . . . 3, 7

*Amborella trichopoda* Baill. . . . . . . . . . . . . . . . . . . . . . . . . . . . . . . . . 1, 3, 5, 6, 7, 8, 9, 10, 11, 12, 13, 17

Amborellaceae . . . . . . . . . . . . . . . . . . . . . . . . . . . . . . . . . . . . . . . . . . . . . . . . . . . . . . . . . . . . . . . . 1, 3, 4

*Hedycarya* J.R.Forst. & G.Forst. . . . . . . . . . . . . . . . . . . . . . . . . . . . . . . . . . . . . . . . . . . . . . . . . . . . . . . 3

Monimiaceae . . . . . . . . . . . . . . . . . . . . . . . . . . . . . . . . . . . . . . . . . . . . . . . . . . . . . . . . . . . . . . . . . 1, 3, 7

*Pythium splendens* Braun. . . . . . . . . . . . . . . . . . . . . . . . . . . . . . . . . . . . . . . . . . . . . . . . . . . . . . . . . . 11

*Tambourissa* Sonn. . . . . . . . . . . . . . . . . . . . . . . . . . . . . . . . . . . . . . . . . . . . . . . . . . . . . . . . . . . . . . . . 7

# THE GLOBAL FLORA

## SERIES 1 – LYCOPODS

Isoëtaceae   Lycopodiaceae   Selaginellaceae

## FERNS

Aspleniaceae   Gleicheniaceae   Marsileaceae   Psilotaceae
Cyatheaceae   Hymenophyllaceae   Matoniaceae   Pteridaceae
Cystodiaceae   Lindsaeaceae   Ophioglossaceae   Saccolomataceae
Dennstaedtiaceae   Lonchitidaceae   Osmundaceae   Salviniaceae
Dipteridaceae   Marattiaceae   Polypodiaceae   Schizaeaceae
Equisetaceae

## GYMNOSPERMS

Araucariaceae   Ephedraceae   Pinaceae   Taxaceae
Cupressaceae   Ginkgoaceae   Podocarpaceae   Welwitschiaceae
Cycadaceae   Gnetaceae   Sciadopityaceae   Zamiaceae

## SERIES 2 – ANGIOSPERMS

Acanthaceae   Asparagaceae   Cabombaceae   Commelinaceae
Achariaceae   Asphodelaceae   Cactaceae   Connaraceae
Achatocarpaceae   Asteliaceae   Calceolariaceae   Convolvulaceae
Acoraceae   Asteraceae   Calophyllaceae   Coriariaceae
Actinidiaceae   Asteropeiaceae   Calycanthaceae   Cornaceae
Adoxaceae   Atherospermataceae   Calyceraceae   Corsiaceae
Aextoxicaceae   Austrobaileyaceae   Campanulaceae   Corynocarpaceae
Aizoaceae   Balanopaceae   Campynemataceae   Costaceae
Akaniaceae   Balanophoraceae   Canellaceae   Crassulaceae
Alismataceae   Balsaminaceae   Cannabaceae   Crossosomataceae
Alseuosmiaceae   Barbeuiaceae   Cannaceae   Crypteroniaceae
Alstroemeriaceae   Barbeyaceae   Capparaceae   Ctenolophonaceae
Altingiaceae   Basellaceae   Caprifoliaceae   Cucurbitaceae
Alzateaceae   Bataceae   Cardiopteridaceae   Cunoniaceae
Amaranthaceae   Begoniaceae   Caricaceae   Curtisiaceae
Amaryllidaceae   Berberidaceae   Carlemanniaceae   Cyclanthaceae
AMBORELLACEAE 3: 1-20   Berberidopsidaceae   Caryocaraceae   Cymodoceaceae
Anacampserotaceae   Betulaceae   Caryophyllaceae   Cynomoriaceae
Anacardiaceae   Biebersteiniaceae   Casuarinaceae   Cyperaceae
Ancistrocladaceae   Bignoniaceae   Celastraceae   Cyrillaceae
Anisophylleaceae   Bixaceae   Centroplacaceae   Cytinaceae
Annonaceae   Blandfordiaceae   Cephalotaceae   Daphniphyllaceae
Aphanopetalaceae   Bonnetiaceae   Ceratophyllaceae   Dasypogonaceae
Aphloiaceae   Boraginaceae   Cercidiphyllaceae   Datiscaceae
Apiaceae   Boryaceae   Chloranthaceae   Degeneriaceae
Apocynaceae   Brassicaceae   Chrysobalanaceae   Diapensiaceae
Apodanthaceae   Bromeliaceae   Circaeasteraceae   Dichapetalaceae
Aponogetonaceae   Brunelliaceae   Cistaceae   Didiereaceae
Aquifoliaceae   Bruniaceae   Cleomaceae   Dilleniaceae
Araceae   Burmanniaceae   Clethraceae   Dioncophyllaceae
Araliaceae   Burseraceae   Clusiaceae   Dioscoreaceae
Arecaceae   Butomaceae   Colchicaceae   Dipentodontaceae
Argophyllaceae   Buxaceae   Columelliaceae   Dipterocarpaceae
Aristolochiaceae   Byblidaceae   Combretaceae   Dirachmaceae

Doryanthaceae
Droseraceae
Drosophyllaceae
Ebenaceae
Ecdeiocoleaceae
Elaeagnaceae
Elaeocarpaceae
Elatinaceae
Emblingiaceae
Ericaceae
Eriocaulaceae
Erythroxylaceae
Escalloniaceae
Eucommiaceae
Euphorbiaceae
Euphroniaceae
Eupomatiaceae
Eupteleaceae
Fabaceae
Fagaceae
Flagellariaceae
Fouquieriaceae
Francoaceae
Frankeniaceae
Garryaceae
Geissolomataceae
Gelsemiaceae
Gentianaceae
Geraniaceae
Gerrardinaceae
Gesneriaceae
Gisekiaceae
Gomortegaceae
Goodeniaceae
Goupiaceae
Griseliniaceae
Grossulariaceae
Grubbiaceae
Guamatelaceae
Gunneraceae
Gyrostemonaceae
Haemodoraceae
Halophytaceae
Haloragaceae
Hamamelidaceae
Hanguanaceae
Heliconiaceae
Helwingiaceae
Hernandiaceae
Himantandraceae
Huaceae
Humiriaceae
Hydatellaceae
Hydrangeaceae
Hydrocharitaceae
Hydroleaceae
Hydrostachyaceae

Hypericaceae
Hypoxidaceae
Icacinaceae
Iridaceae
Irvingiaceae
Iteaceae
Ixioliriaceae
Ixonanthaceae
Joinvilleaceae
Juglandaceae
Juncaceae
Juncaginaceae
Kewaceae
Kirkiaceae
Koeberliniaceae
Krameriaceae
Lacistemataceae
Lamiaceae
Lanariaceae
Lardizabalaceae
Lauraceae
Lecythidaceae
Lentibulariaceae
Lepidobotryaceae
Liliaceae
Limeaceae
Limnanthaceae
Linaceae
Linderniaceae
Loasaceae
Loganiaceae
Lophiocarpaceae
Lophopyxidaceae
Loranthaceae
Lowiaceae
Lythraceae
Macarthuriaceae
Magnoliaceae
Malpighiaceae
Malvaceae
Marantaceae
Marcgraviaceae
Martyniaceae
Maundiaceae
Mayacaceae
Mazaceae
Melanthiaceae
Melastomataceae
Meliaceae
Menispermaceae
Menyanthaceae
Metteniusaceae
Microteaceae
Misodendraceae
Mitrastemonaceae
Molluginaceae
Monimiaceae

Montiaceae
Montiniaceae
Moraceae
Moringaceae
Muntingiaceae
Musaceae
Myodocarpaceae
Myricaceae
Myristicaceae
Myrothamnaceae
Myrtaceae
Nartheciaceae
Nelumbonaceae
Nepenthaceae
Neuradaceae
Nitrariaceae
Nothofagaceae
Nyctaginaceae
Nymphaeaceae
Nyssaceae
Ochnaceae
Olacaceae
Oleaceae
Onagraceae
Oncothecaceae
Opiliaceae
Orchidaceae
Orobanchaceae
Oxalidaceae
Paeoniaceae
Pandaceae
Pandanaceae
Papaveraceae
Paracryphiaceae
Passifloraceae
Paulowniaceae
Pedaliaceae
Penaeaceae
Pennantiaceae
Pentadiplandraceae
Pentaphragmataceae
Pentaphylacaceae
Penthoraceae
Peraceae
Peridiscaceae
PETENAEACEAE 2: 1-16
Petermanniaceae
Petrosaviaceae
Phellinaceae
Philesiaceae
Philydraceae
Phrymaceae
Phyllanthaceae
Phyllonomaceae
Physenaceae
Phytolaccaceae
Picramniaceae

Picrodendraceae
Piperaceae
Pittosporaceae
Plantaginaceae
Platanaceae
Plocospermataceae
Plumbaginaceae
Poaceae
Podostemaceae
Polemoniaceae
Polygalaceae
Polygonaceae
Pontederiaceae
Portulacaceae
Posidoniaceae
Potamogetonaceae
Primulaceae
Proteaceae
Putranjivaceae
Quillajaceae
Rafflesiaceae
Ranunculaceae
Rapateaceae
Resedaceae
Restionaceae
Rhabdodendraceae
Rhamnaceae
Rhizophoraceae
Ripogonaceae
Rivinaceae
Roridulaceae
Rosaceae
Rousseaceae
Rubiaceae
Ruppiaceae
Rutaceae
Sabiaceae
Salicaceae
Salvadoraceae
Santalaceae
Sapindaceae
Sapotaceae
Sarcobataceae
Sarcolaenaceae
Sarraceniaceae
Saururaceae
Saxifragaceae
Scheuchzeriaceae
Schisandraceae
Schlegeliaceae
Schoepfiaceae
Scrophulariaceae
Setchellanthaceae
Simaroubaceae
Simmondsiaceae
Siparunaceae
Sladeniaceae

| | | | |
|---|---|---|---|
| Smilacaceae | Styracaceae | Thurniaceae | Urticaceae |
| Solanaceae | Surianaceae | Thymelaeaceae | Vahliaceae |
| Sphaerosepalaceae | Symplocaceae | Ticodendraceae | Velloziaceae |
| Sphenocleaceae | Talinaceae | Tofieldiaceae | Verbenaceae |
| Stachyuraceae | Tamaricaceae | Torricelliaceae | Violaceae |
| Staphyleaceae | Tapisciaceae | Tovariaceae | Vitaceae |
| Stegnospermataceae | Tecophilaeaceae | Trigoniaceae | Vochysiaceae |
| Stemonaceae | Tetracarpaeaceae | Trimeniaceae | Winteraceae |
| Stemonuraceae | Tetrachondraceae | Triuridaceae | Xeronemataceae |
| Stilbaceae | Tetramelaceae | Trochodendraceae | Xyridaceae |
| Strasburgeriaceae | Tetrameristaceae | Tropaeolaceae | Zingiberaceae |
| Strelitziaceae | Theaceae | Typhaceae | Zosteraceae |
| Stylidiaceae | Thomandersiaceae | Ulmaceae | Zygophyllaceae |

## SERIES 3 – SPECIAL EDITIONS

INTRODUCTION 1: 1-35

www.ingramcontent.com/pod-product-compliance
Lightning Source LLC
Chambersburg PA
CBHW040305220526

45473CB00002B/584